YOUR KNOWLEDGE HAS VALUE

- We will publish your bachelor's and
  master's thesis, essays and papers

- Your own eBook and book -
  sold worldwide in all relevant shops

- Earn money with each sale

Upload your text at www.GRIN.com
and publish for free

Spiral Progression Approach in Teaching Science (SPATS). Enhancing and Developing Science Skills

Niña Espineli

Bibliographic information published by the German National Library:

The German National Library lists this publication in the National Bibliography; detailed bibliographic data are available on the Internet at http://dnb.dnb.de.

ISBN: 9783346835185
This book is also available as an ebook.

© GRIN Publishing GmbH
Nymphenburger Straße 86
80636 München

All rights reserved

Print and binding: Books on Demand GmbH, Norderstedt, Germany
Printed on acid-free paper from responsible sources.

The present work has been carefully prepared. Nevertheless, authors and publishers do not incur liability for the correctness of information, notes, links and advice as well as any printing errors.

GRIN web shop: https://www.grin.com/document/1335116

Teachers Lived Experiences on the Implementation of SPATS (Spiral Progression

Approach in Teaching Science): Basis for Enhancing and Developing

Science Skills in Teaching

Contents

Teachers Lived Experiences on the Implementation of SPATS (Spiral Progression

Approach in Teaching Science): Basis for Enhancing and Developing

Science Skills in Teaching

1. ABSTRACT

Niña P. Espineli

The study was conducted at Trece Martires City National High School - Main, the respondents were 25 secondary science teachers. The main objective of the study was to explain the experiences and challenges encountered by science teachers of TMCNHS in the implementation of Spiral Progression Approach in Teaching. Thematic Analysis was used to analyze the information given by the respondents which was gathered through informal interview and focal group discussion. The study revealed the strengths and weakness of SPATS. In terms of content and pedagogy, Spiral Progression Approach in Teaching Science was found to be learners-centered, advanced, sophisticated, dynamic, productive and innovative. SPATS consider the diversity of learners by matching the lessons to the level of comprehension and through differentiated instruction. Integration of ICT for instruction and continuous use of localized materials enhances the learning of the students. On the side of the teacher, it stimulates intrinsic motivation and upgrade the skills of the teacher.

However, the science teachers experienced the drawback of spiral progression approach despite of commendable results. Challenges encountered by the respondents were classified as to personal and technical factors. Spiral Progression Approach in Teaching requires a Highly Competent Teacher, sufficient materials and facilities.

The data gathered in the study will serve as base line data for the constructive recommendation in accord with the response of the key informants. This research may be useful to other subject areas that also uses spiral progression in teaching. Further studies may focus on pedagogical approaches, the 2C2IR and its impact on subjects delivered through spiral progression approach in teaching.

Keywords:
Spiral Progression Approach, Teachers' Experience, Dynamic, Innovative

The researcher would like to express her profound gratitude to the following people who helped and supported her in the conduct of the study.

Dr. Famie C. Apay, Principal IV of Trece Martires City National High School, for the moral support and encouragement.

Mrs. Marissa M. Rodil, MT II and OIC of TMCHS – Hugo Perez Extension, for her patience and valuable suggestions for the success of the study.

Mrs. Ma. Theresa E. Obrero, MT of Bagong Pook Elementary School for sharing her ideas on how the study will be conducted.

The MRC and SRC for their valuable comments and suggestions.

The Science teachers of TMCNHS for their kindness and word of encouragement.

The K to 12 Program covers Kindergarten and 12 years basic education to provide sufficient time for mastery of concepts and skills, develop lifelong learners, and prepare graduates for tertiary education, middle-level skills development, employment and entrepreneurship.

Spiral progression is an approach used to teach science and mathematics under K to 12 curriculum. Its focus is all about understanding that human cognitive evolved in a step-by-step process of learning, which relied on environmental interaction and experience to form intuition and knowledge. Progression can be vertical, increasing complexity of the lesson or horizontal having broader range of applications. From the study of Angeles (2013) as cited by Resurreccion et.al (2015), the new curriculum is composed of set of activities like collaborative learning, peer tutoring, outcome based performance or performance tasks. Using spiral progression approach, students were exposed to authentic assessment instead of traditional classroom assessment. Authentic assessment means that the task your student will perform should be related/related to the task they are doing in the real world. This includes Project Based Learning, Performance task, Portfolio, Collaborative works and online examinations.

Spiral Progression approach involves vertical and horizontal articulation of competencies. Vertical articulation serves as a bridge of knowledge from one lesson to the next, across a program of study. While, horizontal articulation integrates the skills and knowledge across different discipline. It means that what has been studied in one specific course or area in line with the other.

As mentioned by Montebon (2014) among the different subjects, science is one of the subject that underwent major revisions. In the old curriculum, science subjects (earth science, biology, chemistry and physics) were offered one in each year level and were taught using discipline-based approaches. In the new curriculum, however, science concepts and applications in all subjects are introduced in a spiral approach. In terms of instruction, the science program shifted from traditional methods of teaching to a more innovative exploration that emphasizes the enhancement of the students critical thinking and scientific skills.

In an interview with Dr. Tapang, chair of Agham Advocates of Science and Technology for people, he mentioned that "developing a domestic industry starts with correct track in education. Science is not just the facts and figures we have in textbooks. It is the observational and analytical skills that we gain in these science subjects that make us act and think "scientific" in our lives".

In teaching science, teachers must learn how apply different strategies and techniques that is suited to spiral progression approach. The study of Resurreccion et.al (2015) found that spiral progression approach had greatly influenced science curriculum particularly the content and transitions of four areas in Science. They further explained that science teachers were still adopting to the new curriculum they needed more time and trainings to master all fields and learn new teaching strategies. Likewise, Samala (2017) study revealed that spiral progression provides a deep understanding of science concepts through review conducted by the teacher. It was also emphasized in the study that the use of instructional materials that fit the interests of the students and the mastery of the subject matter of the teacher helped in the retention of science concepts.

Several findings about SPATS (Spiral Progression Approach in Teaching Science) encourages the researcher to conduct a study. This study focused on determining the experiences and challenges encountered by science teachers of

TMCNHS in the implementation of SPAT. The researcher believed that this study provides information/feedback to the curriculum planner on how to continuously improve the program.

Resurreccion et.al (2015) found out that science teachers were still adopting to the new curriculum, they needed more time and trainings to master all the fields and learn new teaching strategies because it is difficult to teach something, in which one does not have the necessary mastery. They can teach other branches of science without the in-depth discussion because it is not their specialization.

From the study of Valin et.al (2019) it was revealed that in vertical articulation, most of the teachers perceived that the topics discussed in the previous grade level are pre-requisites in the present grade level. In terms of horizontal articulation, teachers believed that the concepts and skills in life science, chemistry, physics and earth science are presented with increasing level of complexity from one grade level to another. There are different strategies used by the teachers in the implementation of Spiral Progression Approach in teaching. However, science teachers encountered several difficulties in the implementation of spiral progression approach along teaching strategy and instructional materials.

Orbe (2018) study suggested that the teacher's content, pedagogy, and assessment in chemistry are problematic, specifically, challenges such as instruction-related factors, teachers competence, in-service training sufficiency, job satisfaction, support from upper management, laboratory adequacy, school resources, assessment tools, and others influence teacher success in teaching chemistry. These identified challenges greatly affect the ultimate beneficiaries of education which is the learner.

Science curriculum framework of high performing countries follow a spiral progression, and integrated approach at least up to G9. The curriculum of high performing countries such a Australia, Brunei, England, Finland, Japan, Taiwan, Thailand, Singapore, New Zealand and USA gives emphasis on the connection across topics and disciplines scientific literacy. Based on CEAP National Convention 2019 as cited by Tan (2012)

5. RESEARCH QUESTIONS:

The study was conducted to explain the experiences and challenges encountered by science teachers of TMCNHS in the implementation of Spiral Progression Approach in Teaching. Specifically, this answers the following questions:

1. What are the experiences and challenges encountered by the teachers in the implementation of the spiral progression approach in teaching science subject?

2. What are the proposed action plan after knowing the experiences and challenges encountered by Science teachers in TMCNHS?

3. Based from the result, what is the future direction of the study?

6. RESEARCH PARADIGM

The assessment of different experiences and challenges of Secondary Science teachers in Trece Martires City National High School – Main in connection with the implementation of SPAT (Spiral Progression Approach in Teaching) Science follows the process or paradigm below. This would serve as the basis in giving recommendations on the result of the study.

7. SCOPE AND LIMITATION

The study was conducted at Trece Martires City National High School-Main for the period of two months from July to August 2019. Twenty five (25) Science teachers were the respondents, they are those who are teaching in the above mentioned school for the period of 3 years and above. The study was limited in determining the experiences and challenges encountered by the teachers in using the spiral progression approach in teaching science subject.

8. RESEARCH METHODOLOGY

The study utilized the qualitative phenomenological method of research to describe the experiences of the 25 Science teachers and the challenges of the aforementioned approach in teaching.

9. Sampling

Purposive sampling was utilized in determining the respondents of the study. Twenty five (25) Science teachers of Trece Martires City National High School - Main were the key informants in this study. Out of 25 informants 3 are males and 22 are female. Among the 25 science teachers, 5 are major in Gen. Science, 16 are major in Biology, 3 major in Chemistry and 1 major in Physics.

Profile	Number of Respondents	Percent
Gender		
Male	3	12%
Female	22	88%
Major Field		
Earth Science	5	20%
Biology	16	64%
Chemistry	3	12%
Physics	1	4%
Number of Years in Teaching the major field		
Did not experience teaching the major field of specialization	9	36%
Less than 10 years	5	20%
11 – 20 years	8	32%
21 – 30 years	3	12%

Table 1: Demographic profile of the participants of the study

10. Data Collection

The researcher utilized the FGD (Focal Group Discussion) and interview to seek the in-depth information that will reveal the current situation on how the informants responded to the implementation of approach.

11. Ethical Issues

A permit to conduct the study addressed the administrator of the school was accomplished before the data was collected, this includes the respondents of the study. A request addressed to the informant asking them to participate in the conduct of study was also be provided. Likewise, confidentiality of the respondents' answer was strictly considered.

12. Plan for Data Analysis

In analyzing the collected data, the researcher used of thematic analysis according to Braun and Clarke (2013), thematic analysis is a flexible data analysis that qualitative research use to generate themes from interview data following the 6 phases of thematic analysis.

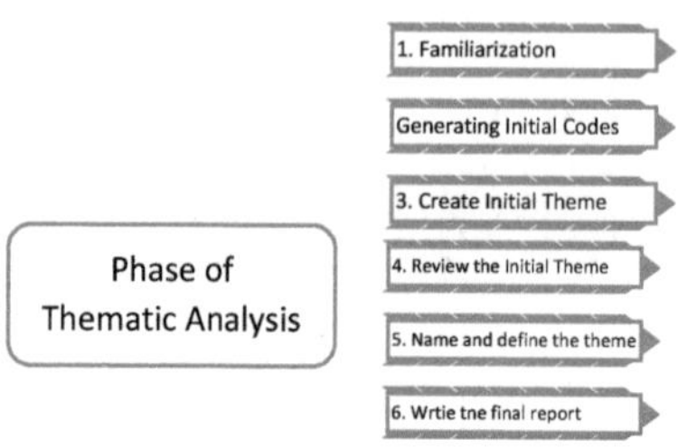

After following the phases of analyzing data, the outcome of the study was presented to the key informants for the validity of their responses and the formulation of the action plan for the development program of Science teachers with regards to the implementation of Spiral Progression Approach in Teaching (SPAT). The researcher asked assistance from the experts (internal and external validator) for the legitimacy of the themes used by the researcher anchored to the responses of the key informants.

13. RESULTS AND DISCUSSION

The findings of the study utilized a qualitative means of investigation. Through focus group discussion and personal interview an in-depth information that unfolds the current situation on how the informants responded to the implementation of using Spiral Progression Approach in Teaching Science.

The shared Science teachers' insights and responses served as the basis for the thematic analysis of the data.

Objective 1. Experiences and challenges encountered by the teachers in the implementation of the spiral progression approach in teaching science subject.

14. Spiral Progression of the Content

As experienced by Science teacher two major characteristics were uncovered. While science teachers viewed Spiral Progression Approach in Teaching as learners-centered, advanced and sophisticated the claim that it is not concentrated, extensive and challenging remain to be its biggest impediments.

15. Spiral Progression of the Content is Learners – centered, advanced and sophisticated

From the FGD and interview it was found that spiral progression of content to be learner-centered. As cited by Ferido (2013) spiral progression approach is not only vertical (increasing complexity), but also horizontal (broader range of application). Science instruction in the K to 12 allows the students to update retained knowledge from four areas of science. Since it is ladderised, each grade level receives a piece of subject area and so updating is made easier. It improves learning since students were given preliminary knowledge in science as they enter high school education for having all these subjects in every year. As shared by the respondents:

Science lessons were easier to understand and mastered because some topics are offered in all grade level. There is a continuity of the lesson since prerequisite were taught in the earlier years.(R1) Lessons in Science allows learners to learn the topics and skills appropriate to their level of understanding. (R21) The spiral progression approach in teaching science minimizes spoon feeding of information to the students. Students' best learned through experience, relating the lesson to a real life situation enable the students' to learn better. (R13) Practicing their skills helps my students increase their knowledge of the lessons being presented. (R4) Developing critical thinking skills among our learners are our top priority, introduction of HOTS questions, exploration and discovering the concepts by themselves and realization of the importance of the lesson is important. (R5)

For our students to become hooked and enthusiastic about the topics being discussed, for them to discover things on their own and from their peers several strategies were applied. (R6)

For me, spiral progression is the strength of the K to 12 since the teachers are encouraged to make efforts to study the four areas of science. (R7) In my case, I really read a lot, I make sure I understand the lesson before introducing it. Since some lessons are introduced each year, I am trying my best to improve what I've been delivered in the past year, either through my power point presentation, challenging activities, learners' focused tasks etc. (R8)

Additionally, with the introduction of learning strategies suited for the learners need, spiral progression approach of teaching leads to a more advanced and sophisticated or complex lessons/ content in science. According to Chapman (2002) as cited by Orbe (2018) in particular, it is felt that when focusing upon progression, it is the nature and sequencing of learning activities that need to be highlighted by those responsible for ensuring progression in schemes of work.

16. Spiral Progression on Pedagogy

In the study, the experiences and challenges of science teacher in the implementation of spiral progression approach in teaching as to pedagogy were explored and two major themes have been identified. While some teachers recognized the learning environment of science instruction as being dynamic, realistic and productive, most of the teachers commented the other way around.

 Spiral Progression in Teaching Science is Dynamic, Innovative and Productive

Under the K to 12 the science instruction focuses on real-life situations and give importance to improvisation and localization. Interest of the students enable them to be more engage in doing activities which will help them better understand the lesson through inquiry based learning. Learning by doing among students

that students learn best. According to Kerner (2018) there are five advantages of experiential learning. This includes: ability to immediately apply knowledge, access to real-time coaching and feedback, promotion of teamwork and feedback, development of reflective practice habits. Interest of the student should be accompanied with teachers' proper guidance and monitoring. According to Barron et.al (2008) as cited by Orbe et.al (2018) "learning by doing" has somewhat checkered track record, in part because teacher often lack information, support and tools necessary to fully integrate and support this alternative approach of teaching and learning. As mentioned by the respondents:

Not all activities in the Learners Materials can be conducted for not all materials can be provided by the school. As a G10 teacher, I try to improvise the activity by using localized materials. This will somehow enhanced the learners understanding towards the lesson presented. (R5) In grade 8 if we can improvised the activity we do it, but if we'll find difficulty we are using video assisted materials like what we downloaded from the several online app provider for our students better understand the lesson. (R1)

In my lesson in grade 7 about identifying acid bases, we make allow our students use an eggplant as indicator since it is readily available in the market and at home. Then after the activity I will let them watch the video which is also the same as the lesson, this to the lesson. (R2)

For me experimentation first before the lecture discussion is still the best strategy to facilitate learning among students. I also used power point presentation and video assisted instruction for the lesson.(R7) As grade 10 teacher, I used videos especially when introducing topics which is difficult for students to understand. Example of which is motor and generator which is very complicated to introduced, but with the utilization of videos imparting lesson about this becomes easier. (R19)

Table 2. Strength of the spiral progression approach in teaching Science.

Responses	Basic Theme	Organizing Themes	Global Themes
1. Students were engaged in relating the lesson to a real life situation	Reality-based learning	Learner Centered	Content And Pedagogy
2. Student learn best through repeated experience and learning of concepts	Participatory learning		
3. The complexity of the lesson matches students readiness and level of comprehension	Differentiated instruction	Advanced and Sophisticated	
4. Do activity by using localized materials.	Localization of Materials	Dynamic way of Teaching	
5. Video assisted instruction	Use of multimedia	Innovative	
6. Practicing the students skills through performance task and presentation	Cooperative & inquiry based learning	Productive	
1. The dedication of the teachers in preparing an upgraded and advanced lessons	Teacher's dedication to work	Intrinsic motivation	Positive Internal Factor
2. Upgrading teachers' skills and ability through continuous studying.	Academic Advancement	Self-encouragement	

Table 3. Weaknesses of the spiral progression approach in teaching Science.

Responses	Basic Theme	Organizing Themes	Global Themes
1. Limited attention span of the student	Student attention span	Students' Concentration	
2. Some of the lessons are not pre-requisite on higher grade		Progression of the lesson	

			Technical Factor
3. Slow learners remain slow, (general curriculum)	Unequal opportunity	Limited opportunity	
4. Not all students mastered the principles and concepts in their lower grade	Mastery of the lesson		
5. The instruction lacks depth and students' retention becomes very slow due to lack of time	Lack of time for time for discussion	Time constraints	
6. Lesson was not finished within the allotted period per day.	Time constraint		
1. I find it difficult to deliver the lesson due to fast transition from one branch of science to another	Difficult	Complicated	Personal Factor
2. Without materials to use, I found difficulty in delivering the lesson clearly	Teacher's reluctant to explore for alternative	Teacher factor	

18. Spiral Progression of the Content is Not concentrated, Extensive and Challenges Instruction

Based on the lived experience of science teachers, they revealed their letdown as they narrated their participation in the spiral progression approach in teaching that is not concentrated, extensive and challenges instruction. While most of the respondents of this study expressed their approval of the ladderised style as new curriculum design in terms of content, many stated due to time constrains the content standard suffers. Since subject is changing quarterly, focus becomes minimal, lack of depth and lack of concentration which is not the goal of spiral progression. Ferido (2013) explained that in spiral progression, the more facts and principles on each topic are encountered, the understanding grows in breadth and depth. As mentioned by some respondents.

A grade 10 teacher said "though lessons were taught previously I think the attention span of the student may be a factor why many of the topics were forgotten by the learners. I need to restart or open again the lesson (review) before introducing a new one". (R1) Sometimes lessons are not pre-requisite on higher grade. Example, mechanics are being taught in g9 while Optics and Magnetism are the focus in grade 10. It sometimes lead to confusion and I experience difficulty to sequence instruction to ensure that learners acquire the necessary pre-skills before introducing a difficult skills. (R6) Due to time constrains for every science subject, the instruction lacks depth and students' retention becomes very slow. (R10) Not all students mastered the principles and concepts in their lower grade. Example will be the problem solving lessons in Chemistry which requires skills in converting measurement from one unit to another. Students who do not have skills find difficulty in solving problems. (R4) Since, we have limited materials, we usually do group activities but not all are productively participating. Many are just relying on the work of their leaders, slow learners tend to sit back and copy the answers of their classmates. (R16) Since SPATS focused on learners and lots of group activities were presented often times lessons were not finished. Sometimes it takes three days to complete a single lesson. (R5)

The spiral progression approach in teaching science in terms of content challenges instruction since it tests teachers' self-efficacy and competence. Teachers cannot teach what they do not know, if teachers are not knowledgeable of the lesson, not well trained and lack of the necessary preparation to use spiral progression approach then, delivery of learning to the student will be a great problem. Spiral progression requires heavy preparation since every teacher must learn and teach the four science subjects. Teaching science subject which is not their specialization makes it difficult for them to deliver lessons efficiently.

I am a Biology major, a grade 8 teacher said, teaching Physics for me is very difficult. Peer teaching really helps me a lot, I usually ask my co-teacher that is specialize on the field to help me clarify concepts which I find difficulty. (R1) A grade 10 teacher said "I read a lot, I make sure that I understand the lesson well before introducing it and since some lessons are introduced each year, I try my best to improve what I've been delivered in the past year, either through my presentation, challenging activities, learners focused tasks etc." (R2) Teaching other subject other than my area of specialization forces me to learn concept outside chemistry, it's also self-fulfilling to be able to teach lessons you just learned and

learned in a hard way and as time progress they became parts of my expertise as well. (R3)

Another problem encountered in adopting spiral progression approach is the insufficient laboratory materials and instructional materials such as modules, and books which are required to facilitate learning. As mentioned by the respondent:

Some instructional materials are not available for both teachers and students. Grade 7 lessons in the first quarter do not have materials in the laboratory. (R1) Some activities in the Learners Materials were not performed if we fail to localized the materials since we only have limited materials in the science laboratory. (R2)

19. Spiral Progression Approach in teaching requires a Highly Competent Teacher, and sufficient materials and facilities.

Science teachers are not all science majors. To deliver science instruction a teacher must be competent and qualified. There are some teachers teaching science subject which is not their major of specialization. Since no enough seminars were given to the non-majors, they often rely to good textbook and other innovative references since they need to upgrade their learning for them to deliver instruction to the learners with maximum efficiency. Ideally, teacher should take the responsibility of continuously upgrading their knowledge on content and scientific skills needed for improved instruction. As some science teachers mentioned:

At first, I find it difficult to teach topics or lessons which are not my major of specialization, but as time passed by, I need to be effective teacher so I seek help from my co-teachers. Through peer teaching and coaching, I was able to teach the topic with ease and confidence. (R3) I try to be better in my teaching though it's not my major field, I make sure that every details of the lesson was delivered well. This is by referring to different resources like books and relying to technology to confirm some concepts which I found difficult to explain. Video instruction is really an effective agent to learn. (R) My intention to pursue a masteral degree which in line with the subject (s) will help me improved in terms of content that I am teaching aside from it will give me lots of opportunities afterwards.(R4) Teaching Physics for 20 years and switching it to earth science, biology and chemistry is not an easy task. I need to strive learning again to be somehow efficient teacher in different field, though they are saying that sometimes its good to move away from our comfort zone, still I found difficulty in the transition of what I am teaching.(R19)

Objective No. 2. Proposed Action Plan for teachers on Spiral Progression Approach in teaching Science.

Thereafter knowing the experiences and challenges encountered by the Science teachers in the implementation of Spiral Progression Approach in Teaching (SPAT), several recommendations were stipulated on how to enhance the skills Science teachers in the implementation of SPAT.

Table 4: Action Plan on SPAT Development Plan for Science Teachers

Objectives	Strategies/Activities	Person's Involved	Time Frame	Source of Fund	Expected Output
Enhance the awareness of teachers on SPATS	Have a Focus Group Discussion to all Science teacher of TMCNHS	25 science teachers	September 2019	Science Department Fund	Intensified plan for an in-service for science teachers
Upgrade the knowledge of the teachers about the strategies used to deliver the lessons in increasing difficulty	Peer teaching In-service training on the Development of pedagogical approaches suited to the increasing difficulty of the lesson as part of SPAT	25 science teachers Invited lecturers Department Head School Principal	2 Saturdays of October 2019	MOOE	School based in-service training for Science teachers related to - approaches that will enhance SPAT - Differentiated instruction

Produce an adequate activities in administering Roll-Out on the Development plan on how to deliver SPAT	Administer a district roll – out on the development plan on 1. proper approaches to deliver SPATS 2. Differentiated Instruction	Science teachers in the district of Trece Martires Lecturers (Master Teacher) School heads	October 21-23, 2019 (Semestral Break)	Contribution of each school in the district of Trece Martires	75% of the teacher participants enhanced their knowledge on - approaches that will enhance SPAT - Differentiated instruction

Objective No. 3. The future direction of the study.

The data gathered in the study will serve as base line data for the constructive recommendation in accord with the response of the key informants. Formulation on the development plan of the science department focusing on the approaches to be used to uplift the techniques in teaching science through spiral progression approach was the output of the study. This research may be useful to other subject areas that also uses spiral progression in teaching. Further studies may focus on pedagogical approaches, the 2C2IR and its impact on subjects delivered through spiral progression approach in teaching.

20. Conclusion and Recommendation

As reflected on the results of the participants' responses on the lived experience of the Science Teachers in Trece Martires City National High School-Main showed the advantages and disadvantages of spiral progression approach in teaching science. The result were analyzed using thematic analysis resulting to the formulation of themes anchored to the respondents' responses. The responses were analyzed as to positive and negative impact of spiral progression approach in teaching science as experienced by science teachers. In positive results themes were categorized in content and pedagogy and positive internal factor while the negative outcome, themes were classified as Technical and Personal factor.

Almost all the responses of the respondents agreed that spiral progression approach in teaching science in terms of content and pedagogy to be learners-centered, advanced, sophisticated, dynamic, productive and innovative. This allows the learners to learn explore new ideas by relating the lessons to a rea life situation, undergo repeated experience. SPATS consider the diversity of learners by matching the lessons to the level of comprehension and through differentiated instruction. Integration of ICT for instruction and continuous use of localized materials enhances the learning of the students. On the side of the teacher, it stimulates intrinsic motivation and upgrade the skills of the teacher.

However, the science teachers experienced the drawback of spiral progression approach despite of commendable results. Challenges encountered by the respondents were classified as to personal and technical factors. Millennial learners nowadays have limited attention span, diverse and the curriculum itself is somehow lacking for not all the lesson are pre-requisite in the higher year. Referring to different experiences gathered through FGD, an action plan was made for the experts, ex. master teachers teach the science teachers of different pedagogical approaches suited to the level of learners that enhances the delivery of the lesson through SPATS. The school will also be opened to tapped stakeholders, a request/provide laboratory materials that will help to make our teaching strategy

Through the result of this study, TMCNHS or even DepEd-Cavite may utilize the results in designing various Professional Development Programs/trainings that would uplift the science teachers' capability in choosing the appropriate strategy which is suited to spiral progression in teaching science. This study can also be used to raise questions about the instruction of other discipline that also implement spiral progression approach in teaching like Mathematics.

22

Assistance from the school before and after the conduct of the research were secured. The study was presented to the District Research Proposal colloquium. After the conduct of the study, it will be presented on the Division Research Colloquium and other local conferences. The researcher also plan to seek opportunity for the publication of the research study for further dissemination.

22. REFERENCES:

Kerner, R. 2018. Five Advantages of Experiential Learning. Newsroom, Insights. Retrieved from northwell.edu/news

Resurreccion, J. 2015. SPATS in Selected Private and Public Schools in Cavite Retrieved from: https://www.dlsu.edu.ph

Montebon, D. 2014. K-12 Science Program in the Philippines: Students Perception on Its Implementation. International Journal of Education and Research. Retrieved from https://doi.org.com

Orbe, J et.al. 2015. Teaching Chemistry in a Spiral Progression Approach: Lesson form Science Teachers in the Philippines. Australian Journal of Teacher Education Department. Retrieved from http://ro.ecu.edu.au/ajte/vol43/iss4/2

Samala, H. 2017. Spiral Progression Approach in Teaching: A case study. Retrieved from http://knepublishing.com/index

Tan, M. 2012. Spiral Progression Approach to Teaching and Learning. Retrieved from https://ceap.org.oh

Tapang, G. 2012. Science and K – 12. Retrieved from https://opinion.inquirer.net/22527/science-and-k12.Teachershouldknow

Valin, E. 2019. Spiral Progression Approach in Teaching Science. International Journal of Engineering Science and Computing. Retrieved from https://ijesc.org/vol9/iss3

The table show the computed expenditures during the entire conduct of this study.

Particulars	Amount per unit	Quantity	Total Price
Bond paper	PhP260	1 rim	260
Ink	Php400.00	2 pcs	800
		Total	**PhP1060.00**

24. ANNEX

Annex A. Questionnaire Guide for Interview

Name: _____________________   Area of Specialization: _______________
No. of years teaching: _______________________________________
No. of years teaching the major field of specialization: ___________

Answer the following with honesty, giving examples will be highly appreciated.
Please include the benefits and drawbacks of SPATS.

SPIRAL PROGRESSION
1. Based on experience, how do you explain spiral progression in terms of content
    you are teaching?

    ____
    ____
    ____
    ____
    ____
    ____
    ____
    ____
    ____
    ____

2. How is the way of teaching science using spiral progression approach differ
    from the way Science subject is being taught in the BEC curriculum? Which
    one do you prefer? why?

    ____
    ____
    ____
    ____
    ____
    ____
    ____
    ____
    ____
    ____
    ____
    ____

3. Spiral progression approach in teaching involves deepening the way the
    lessons are being taught. As Science teacher, what strategies do you use to
    increase the level of difficulty in the subject that you are teaching?

    ____
    ____
    ____
    ____
    ____

Do you find any difficulty in applying your chosen strategy (ies)? Please elaborate your answer by giving various examples

Are the strategies applied to increase the level of difficulty in terms of content/lesson being taught always come up with a positive result? Why or why not?

4. As teacher teaching a non-major field, what ways/strategies do you do/apply to deliver the lesson efficiently?

Do you experience difficulty in teaching the subject which is not your major of specialization? Kindly give examples.

5. Can you consider the ladderised instruction in Science helpful to students as they progress from one level to another?

____

____

____

____

____

____

____

____

6. What difficulty does your student experience in the application of spiral progession approach in teaching science?_______________________________

____

____

____

____

____

____

____

____

____

7. Are the lessons being taught in the earlier grade level a pre-requisite in the higher grade level?_______________________________________

____

____

____

____

____

____

____

Give example of the lesson you are currently teaching which you think were not taught in the previous years. _________________________________

____

____

____

As a science teacher, what do you do for your learners to cope up with the lessons you are teaching specially if they don't have enough background regarding the lesson? _____________________________________

____

____

____

8. Is spiral progression approach in teaching science upgrade the quality of learnings learned by the students? Why or why not? _______________________

____

____

____

____

____

____

9. What suggestions can you give to better teach science subject using spiral progression approach in teaching?_______________________________

____

____

____

____

____

____

____

10. **What pedagogical approach do you usually use to deliver authentic learning among your learners?** _______________________________________

____

____

____

____

____

____

11. **Group activity as part of performance –based learning is a strategy where Science teachers are very much use to. What are the advantage and disadvantage of using this strategy?**

____

____

____

____

12. How important to a science teacher, the knowledge of using the appropriate
pedagogical approach in every lesson to be taught?_______________________

13. Being a teacher teaching a not major field, what do you do to improve your knowledge
in terms of pedagogical approach in teaching the subject? _______

YOUR KNOWLEDGE HAS VALUE

- We will publish your bachelor's and master's thesis, essays and papers

- Your own eBook and book - sold worldwide in all relevant shops

- Earn money with each sale

Upload your text at www.GRIN.com and publish for free